DATE DUE			
JAN 0 4 '93	SEP 10 '01		
MAR 1 6 '95	NOV 03 01		
NOV 02 '9	JUN 12		
FEB 1 8 '9	AR 1 0 200		
MAR 04 '97			
MAR 18 9			
JUL 0 8 '97			
AUG 16 '97			
FEB. 13. 1998			
MAY 15 '00			
AUG 1 8 01			
GAYLORD			PRINTED IN U.S.A.

Watching Them Grow

Inside a Zoo Nursery

Text by Joan Hewett

Photographs by Richard Hewett

Little, Brown and Company
Boston   Toronto

Books by Joan and Richard Hewett

The Mouse and the Elephant

Watching Them Grow:
Inside a Zoo Nursery

FIRST EDITION

Library of Congress Cataloging in Publication Data

Hewett, Joan.
  Watching them grow: Inside a zoo nursery

  SUMMARY: A behind-the-scenes peek into the baby animal nursery at the San Diego Zoo.
  1. Zoo animals—Juvenile literature. 2. Animals, Infancy of—Juvenile literature. 3. San Diego, Calif. Zoological Garden—Juvenile literature. [1. Zoo animals. 2. Animals—Infancy. 3. San Diego, Calif. Zoological Garden] I. Hewett, Richard. II. Title.
QL77.5.H48   636.08'899   79-13345
ISBN 0-316-35968-8

BP

*Published simultaneously in Canada
by Little, Brown & Company (Canada) Limited*

PRINTED IN THE UNITED STATES OF AMERICA

Lisa

Kalind

Lock

Drainola

A-Seven

November 12

It's 6:30 P.M. The zoo is closed. The thousands of people who visit the zoo each day have left. The keepers have finished feeding the animals; they too are gone. The loud *elp elp* call of the peacocks breaks the silence, and an occasional roar from one of the tigers or lions carries clearly through the cool night air.

Over in the children's zoo there's still some cooing and

honking, and the faint sound of flapping wings. The doves and ducks haven't quite settled down. It's very dark, but there's a light on in the zoo nursery.

Inside the cheerful yellow nursery Loretta Owen is on duty. Loretta is giving Kalind, a seven-month-old pygmy chimpanzee, his bottle.

Loretta talks to the little pygmy chimpanzee, telling him how cute he is, how smart, and that she knows he's a mischief-maker, but she loves him anyway. Kalind enjoys the sound of her voice, her touch, and best of all he enjoys being the center of attention. He drinks some more of his formula, then looks slowly around. His gaze fixes on the lights, on a hanging mobile, on a toy on the floor. He starts climbing down to get it, changes his mind, scurries back up, and makes a big, brave face at Loretta. "Oh, scaredy cat, scaredy. What are you afraid of? I didn't tell you you couldn't get down," Loretta says. Then she looks at his face. "You are tough, BooBoo," she reassures him, using the nickname she called him when he was just an infant.

Loretta has been a San Diego Zoo nursery attendant for five years. Wild animal babies who are abandoned by their mothers, or whose mothers aren't caring for them properly, are brought up in the nursery. So are animal babies who have physical problems and need special care. Loretta is one of the zoo's four attendants who help "mother" these baby animals. The animals receive almost round-the-clock attention, for each attendant works a separate shift. Loretta's shift is from 3:30 in the afternoon till midnight.

Sometimes there are only one or two baby animals in the nursery, at other times there are as many as six or seven. A deserted baby animal is usually brought into the nursery by the keeper who has found it. No one knows exactly why wild animals in zoos sometimes abandon their own healthy, newborn babies. It is only known that living in captivity can change an animal's basic mothering instinct, that inherited, unthought-out urge to take care of the young. But even when a wild animal mother is behaving normally, she won't always care for her young. An animal mother often will have nothing to do with a baby that is weak, or has a physical defect such as a missing finger. The infant might not survive in the wild, so the mother instinctively rejects it.

"In the wild an abandoned infant would almost certainly die, but at the zoo, we bring up the baby in the nursery," Loretta says with pride.

Every kind of ape baby — gorilla, gibbon, siamang, orang-utan, chimpanzee, and pygmy chimpanzee — has been cared for in the nursery. Apes often catch cold and flu from human beings. Both sicknesses in a baby ape can easily turn into pneumonia and be very serious. When a baby ape develops a cold he or she is promptly treated by a zoo veterinarian, and the baby often spends weeks recuperating in the nursery. Some baby apes, like Kalind, who seem particularly prone to colds and flu, are raised in the nursery till they are one or two years old and do not catch cold so easily.

"There's nothing like having baby apes in the nursery," Loretta says. "They're so appealing . . . affectionate, full of fun and curiosity. They always get themselves into all sorts of trouble." Loretta gives Kalind a hug, puts him down in his playpen, and adds, "It's probably impossible not to love a pygmy chimp."

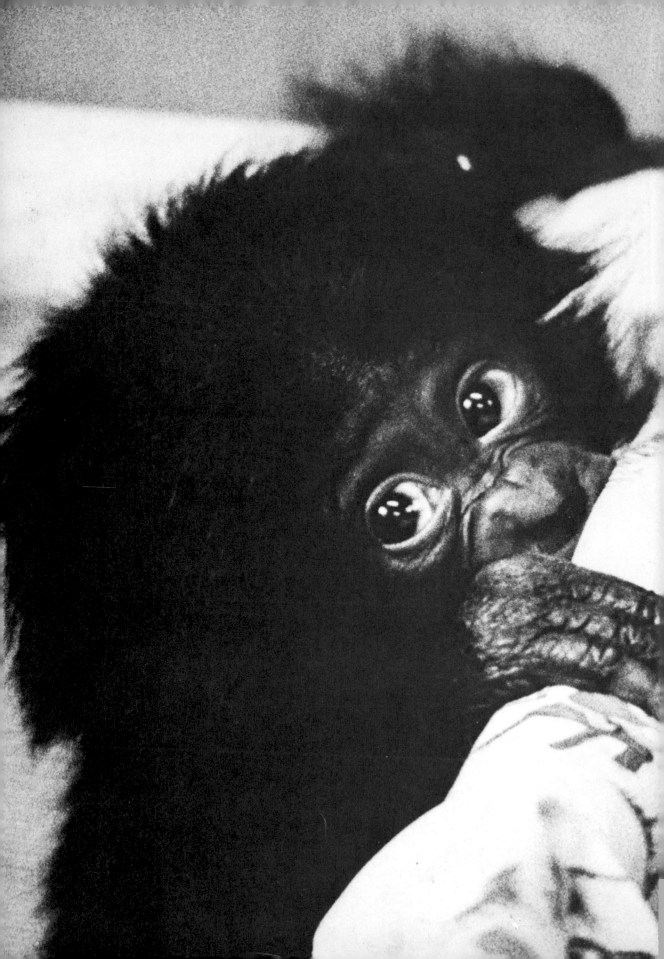

Not many people have ever seen a pygmy chimpanzee. Dark-faced and slender, they are about two-thirds the size of the common chimpanzee, to whom they're related. Remote forest regions in Central Africa are the home of the pygmy chimps. There is now a law protecting them and making it illegal to remove them from Africa. Kalind's parents were brought out long before this law was enacted. Kalind, his brother, and his sisters were all born in the San Diego Zoo.

Counting those in both zoos and research centers, there are fewer than thirty pygmy chimpanzees in captivity in all the world. Scientists are studying them. They've discovered that of all the great apes — gorillas, orangutans, chimpanzees, and pygmy chimpanzees — it is the pygmy chimp who has a brain that may be most like that of primitive man. Kalind and his kin are probably man's closest living relatives.

The nursery is Kalind's home. He resents the two newcomers who are living in his nursery, taking up his "mother's" time. Hanging onto the side of his playpen, his soft black fur bristling with jealousy, he watches Loretta care for the seven-week-old orangutan twins, Lock and Lisa. Lock and Lisa are an event. They are the first twin apes ever born in the zoo.

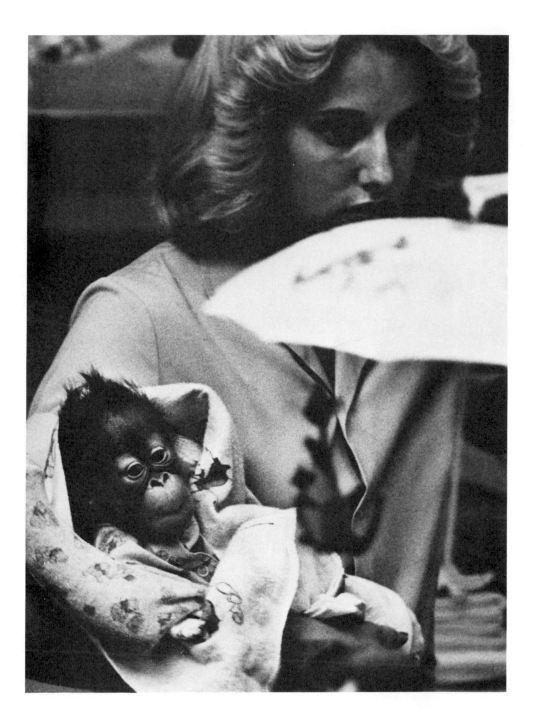

Although they've been in the nursery since they were six hours old, they are not aware of Kalind. Like human infants of the same age, their eyes are just beginning to focus. They notice only those things nearby: Loretta's hands, her face . . . a rattle that she waves in front of them. Loretta fondles and talks to the twins. She gives them their bottles every three hours, changes their diapers, and turns them over from their stomachs to their backs so they won't get uncomfortable in their cribs. Sometimes, while lying in their cribs, one of the orangutans makes a happy, almost gurgling *ga ga* sound.

An ape mother does not instinctively know how to take care of her newborn infant. In the wild she learns by watching her own mother; in captivity she may be able to learn by watching other captive mothers. But the veterinarian who had observed the twins' birth decided that their mother was just too confused to care for them properly. So Lisa and Lock, their coppery orange fur still damp, were rushed to the nursery.

Loretta and JoAnn Thomas, the head nursery attendant, who had stayed to help, took over. They bathed the orangutans in a very warm, gentle sterile solution, dried them thoroughly, and tied their umbilical cords. Then dressing them in hand- and foot-covering stretch sleepers, they placed the twins in 85-degree incubators till they were thoroughly warmed.

Everything that can be done is done to make the shining, clean nursery a safe place for little apes. The nursery temperature is kept at a tropically warm 78 to 80 degrees. Both nursery attendants and veterinarians walk through a shallow plastic pan filled with an antiseptic solution before they enter the nursery, so they won't carry germs in on the soles of their shoes. The attendants keep a detailed record of the animals' progress, and a veterinarian looks in on them each day, and examines them if there are any problems. Visitors watch the babies through picture windows. They are not allowed in the nursery building. Just in case some zoo employee might not realize that the no-visiting rule applies to him, there's a sign on the inner nursery door that warns, NO ONE IS ALLOWED IN. Below that is printed, *"The twins thank you."*

Lisa and Lock are completely helpless in the nursery, but they wouldn't be in the wild. Orangutans are born with a powerful clinging instinct and powerfully strong arms and hands to cling with. In the wild, orangs spend most of their time living in treetops, twenty to sixty feet above the jungle floor. An infant with its arms wrapped about its mother's stomach, its fingers and nails gripping her long hair, remains attached to her while she swings through the trees of the tropical rain forest foraging for food. The orang mother can forage while her baby naps, for the infant does not loosen its grip even while it sleeps.

Without their natural mother to cling to, Lock and Lisa's arms flail about helplessly. So Loretta will mitten the infants' hands as soon as she finishes changing and dressing them. If their hands were not covered, they might cut themselves badly with their pointed, sharp nails. When the babies are just a little older and can control their arms, their hands will be uncovered.

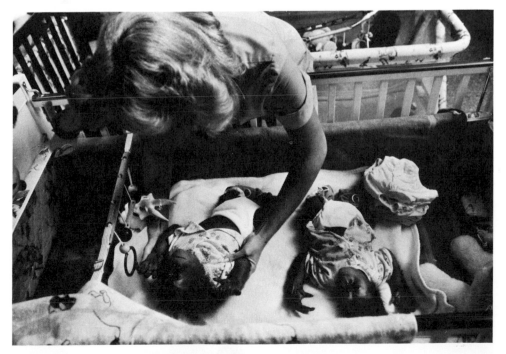

November 19

Loretta says, "Kalind would surely pout, if he knew how." Seven-and-a-half-month-old Kalind is stubborn. He lets Loretta know he doesn't have to do what she asks: hold still, or finish his bottle. He has a "maybe I will, maybe I won't" attitude. Perhaps he's jealous of the twins and wants more attention, or perhaps it's just time for Kalind to see how much his "mother" will let him get away with. Baby pygmy chimpanzees continually practice adult chimp attitudes, gestures, and facial expressions. They make play faces. But whenever Kalind bares his teeth aggressively, like an adult chimp, he suddenly stops playing and looks at Loretta with a worried "are you angry at me?" expression. Loretta gives him a big "I still love you" hug.

Baby chimpanzees, like all the great apes, are strongly attached to their mothers. In the wild, for the first few years, Kalind would be completely dependent on his mother for food, protection, comfort, and love. In the nursery, attendants are his substitute mothers. They fill the pygmy chimp's deep need for mother love, and he responds with great affection.

But raising baby apes is not just a matter of love and kisses. Apes are very intelligent animals; not all their behavior is instinctive. Much of it is learned by watching and imitating. An ape that is brought up by a human family, as a member of the family, imitates his "parents," his "brothers" and "sisters." He believes that he is a person also. By the time the baby ape is three or four years old he is too big and strong to remain with his human family. If the baby is placed in a zoo he is confused and often terrorized by the other apes.

In the nursery the attendants try, as much as possible, to raise the baby apes as apes. Loretta expects Kalind to obey the nursery rules: not to throw his bottle, bite on the nursery furniture, or tear his stuffed animals. He is also not allowed to do anything that will cause him, the other baby animals, or the nursery attendants any harm. But Kalind is never scolded for an instinctive ape response, like making threatening sounds or gestures at a stranger. Loretta tries to help him progress naturally and to develop all the pygmy chimp skills that he should be developing. And, even though he would be a funny, cute mimic, she never ever encourages him to mimic her.

November 26

Today is a banner day! Lock and Lisa spent the whole day without their mittens on. It's an important step for the twins; it's time for them to start discovering their hands and their feet.

"When an orangutan baby grips hold of you, pinching your skin, you really know you've been pinched," Loretta says. "We have to teach them not to hang on when they're held. Lock is very trusting, he rarely grabs me, but Lisa is not that confident. She wants to follow her instinct and hang on. When I undo her grip, she tenses, arches her back, and flails her arms about just a mite."

Nine-week-old baby orangutans in their jungle homeland would still be clinging to their mothers, receiving the direct heat from her body. So even though the nursery is very warm, the twins are clothed for additional warmth. Loretta finishes dressing Lisa. She lays her down in her crib on her stomach and tucks a baby blanket firmly about her so she'll feel secure. Then, to keep the little orang from feeling lonely, she puts a yellow stuffed rabbit near one side of her head and a red and white whale near the other. Picking up Lisa's almost empty bottle she goes over to the record book and writes: "8:30 P.M. Lisa drank four ounces of formula."

The twins are such very dependent little creatures; they still cannot roll over from their stomachs to their backs, and their heads begin to wobble if they hold them up for any length of time. It's almost hard to believe that they really will grow to be great, long-haired orangutans.

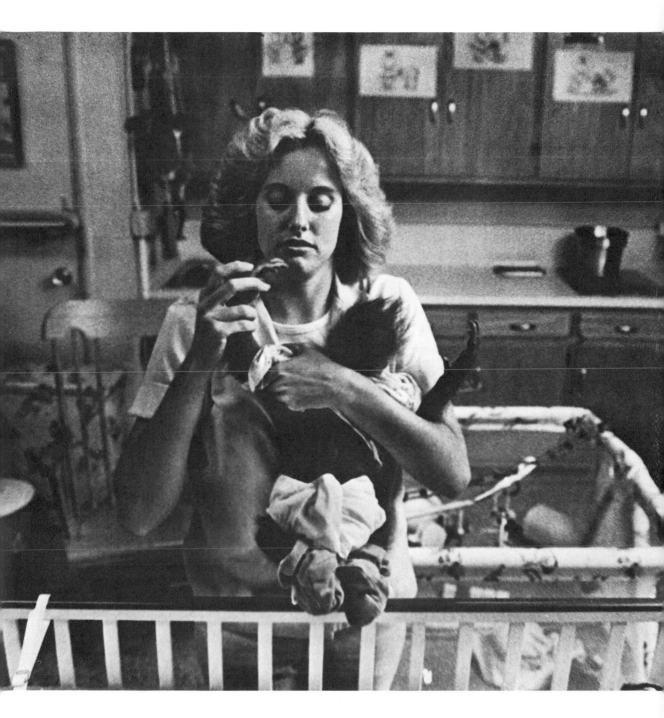

Orangutans come from the Far East, from the tropical rain forests of Borneo and Sumatra. In the language of Borneo, *orangoutang* means "man of the woods." They are very solitary apes. A male orangutan spends most of his time alone and sometimes goes days without seeing another orang. A mother orang travels through the forest with her baby and often an older child. Orangs do not band together in social or family groups, like the other apes. Traveling alone, high up in the treetops, they are so hard to spot that scientists tracking them down are often startled by a first encounter. How mysterious these apes must have seemed to the people of the forests! No wonder that stories and myths about them abound.

There are stories that claim orangs really know how to talk, but they don't want man to know it for they are afraid they'll be put to work. There are stories about male orangutans carrying off beautiful young women. An ancient Asian myth says that two birdlike beings created all forms of life. One day they made a man and a woman. This indeed called for a celebration and the creators feasted and toasted late into the night. The next day they felt quite sick. They tried hard to make more of those wonderful human beings they'd made earlier, but they kept leaving out part of the recipe, and they created orangutans instead.

Unfortunately, there are far fewer orangutans in the wild than there used to be. Hundreds have been captured and sold, many dying from improper care, and giant logging machines and tractors have cleared much of their jungle homeland.

18

Like the rare pygmy chimpanzee, they are now considered an endangered species, and laws prohibiting their capture are strictly enforced.

 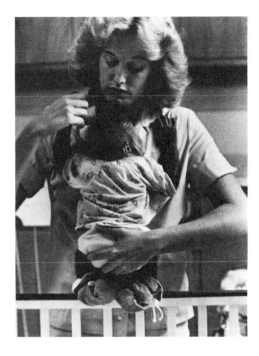

December 2

A tiny spider monkey is the nursery's newest member. Her name is Drainola and she's three days old.

An animal keeper, making his early morning rounds three days ago, spotted a small, furry object lying on the floor of the spider monkey enclosure. A mother spider monkey had rejected her own newborn baby. Its almost lifeless body was curled over the drain. Realizing instantly what had happened, the keeper picked up the infant and, holding her close to him, rushed her over to the nursery.

The one-pound, one-ounce spider monkey was still very weak when Loretta took over at 3:30 P.M. Like many infants who are separated from their mothers, Drainola would not immediately accept or suck from a bottle's nipple. But after several feedings, she seemed to get the hang of it, and by eve-

ning she was drinking willingly. Drainola gets fed round-the-clock, every two hours. When Loretta left at midnight, the late-night attendant continued the feedings until the morning attendant took over at 6:30 A.M.

Three-day-old Drainola now seems calm and alert. Her eyes are wide open and she looks around the incubator with what appears to be great curiosity. Loretta rubs her own hands together, getting them really warm, then lifts Drainola out of her incubator, wraps her in a flannel baby blanket, and feeds her. Drainola looks up at Loretta, till her eyes get heavy with sleep.

Spider monkeys are very common. In the forest of tropical Central and South America, sometimes as many as two hundred spider monkeys live in one square mile. There are many kinds. Some have five fingers, but their fifth finger, their

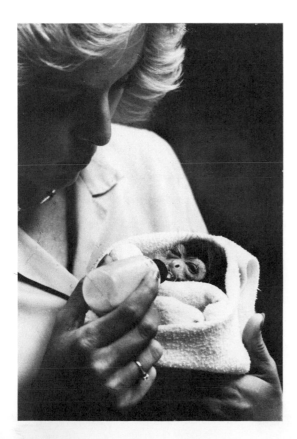 

thumb, is just a stump. Like Drainola, most golden spider monkeys have no thumbs at all. All spider monkeys have extraordinary tails that will wrap around and hold. They are called prehensile tails. When she's older, Drainola will be able to hook her prehensile tail around branches and swing for long periods of time. She'll use her tail for picking up buds and nuts; it will be like an extra hand.

Feeling a slight draft in the nursery, Loretta gets a terry-cloth towel and places it protectively over one side of the incubator. She can't stop worrying about the infant. Is she all right? The veterinarians say so. But still, her mother could have rejected her because something is wrong. Loretta picks up the record book. She reads all the notations that have been made about Drainola since she arrived. Her temperature's been normal, she hasn't been cranky, she's finished most of her bottles, she's gained weight. She's been making very good progress.

The nursery is crammed full of all sorts of equipment to care for the wild animal babies, but the record book is the very center of the nursery system. It helps the veterinarians find out what the normal growth and development is for the different kinds of baby animals, and it is a guide for caring for future nursery animals. The information in the book does not stay within the nursery walls. It's available to any zoo that can use it, for zoos care about their animals and so exchange information freely.

It's 11:15 P.M. Drainola is sound asleep. Loretta looks down at the tiny creature and smiles.

22

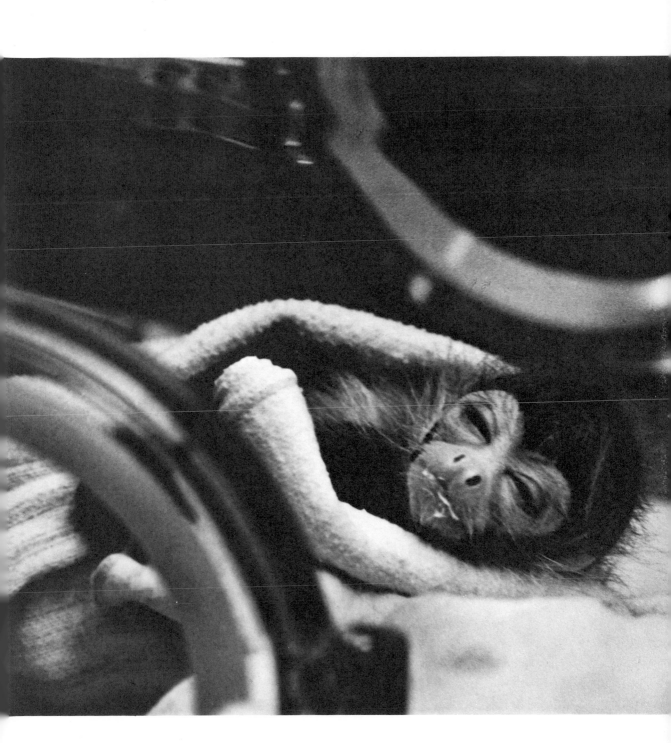

December 12

Two-week-old Drainola has graduated from her incubator to a big crib. She still gets fed every two hours round-the-clock. Kalind seems to have mixed feelings toward Drainola. He doesn't positively resent her — in fact he's quite curious. But sometimes after watching Loretta fuss over the tiny spider monkey for a long time, he'll turn away, drop to the bottom of his playpen, and clutching his blanket for comfort, he'll rock back and forth making a *humf, humf, humf* sound.

Kalind makes a great variety of chirping, chattering, barking, grunting, giggling, yelling, and screaming sounds. Most of these sounds have more than one meaning. Loretta says, "Apes communicate with their bodies as well as their voices. I can tell when a baby ape is in trouble just by listening to its screams, but sometimes I have to see the animal and under-

24

stand what the situation is before I can tell what it wants. When Kalind *humf humf*s, he isn't always feeling sorry for himself, but when he makes the sound softly, holds onto his blanket, and rocks — all at the same time — he's a very sorry-for-himself baby."

When pygmy chimpanzees are warning someone that they want him to go away, their hair stands on end, they show all their teeth, look as if they're about to charge, and make a loud *hee hee* sound. But without the fierce gestures, a rapid, not-so-loud *hee hee*ing can mean happy excitement. Kalind *hee hee*s menacingly once in a while, but not very loudly and not for very long, for an eight-month-old pygmy chimp is still very much a baby.

In the wild Kalind would probably begin to eat solid foods, mainly fruits and nuts, when he was about two years old, and would continue nursing (drinking his mother's milk) till he was at least three. In the nursery baby apes start to get cereal and fruits, along with their formula, when they're seven or eight months old. "They just never like their solids to begin with," Loretta says. "And they're all so smart, it takes them no time to realize that when their bowl is empty, they'll get their bottle. So given a chance, a baby ape will hide his food. Down the diapers," she adds, " is a popular spot." Kalind clamps his mouth shut as soon as the spoon filled with cereal nears his mouth. Finally after much coaxing he opens his mouth, but then like a naughty child, he lets the cereal ooze out. "No," Loretta says, placing a finger firmly but gently on his nose. Kalind's big eyes look at her reproachfully . . . what could he possibly have done? Then he eats correctly for a couple of mouthfuls.

Sometimes when Kalind's finished eating he'll scamper about the nursery floor. It's a big adventure. Eight-month-old Kalind isn't quite old enough to walk steadily on all fours

with straight arms and legs, like an adult chimpanzee. He has his own form of locomotion; he sits on his heels, bends forward, and with his hands on the floor, he scoots about. He's careful, however, never to get more than one or two scoots away from Loretta, and any unexpected noise sends him scurrying back. It takes baby pygmy chimpanzees a long time to feel safe on their own.

Loretta has mothered two other orangutan babies, but "Lock and Lisa," she says, "are my first twins ever."

The two-and-a-half-month-old twins can now lift up their heads when they're on their stomachs and turn themselves over. They like to spend most of their time on their backs, gazing up at their fingers.

Small plastic toys and rattles intrigue them. But trying to grasp with their hands what their eyes can see isn't always easy. Lock has found his jungle gym. He lies and stares up at the dangling objects. If he's too far away, he'll manipulate both his arms and legs with typical orangutan perseverance till he's in a better position. Hooking the shiny plastic ring is difficult; it takes Lock's complete concentration. He reaches for it several times before he's able to touch it. Then slowly, carefully, he manipulates his thumb and four fingers — he has it!

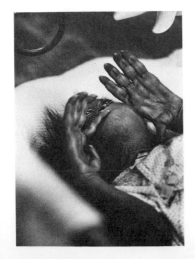 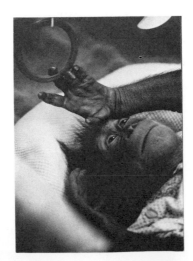 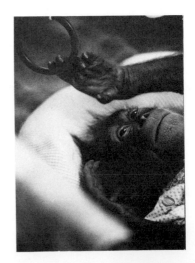

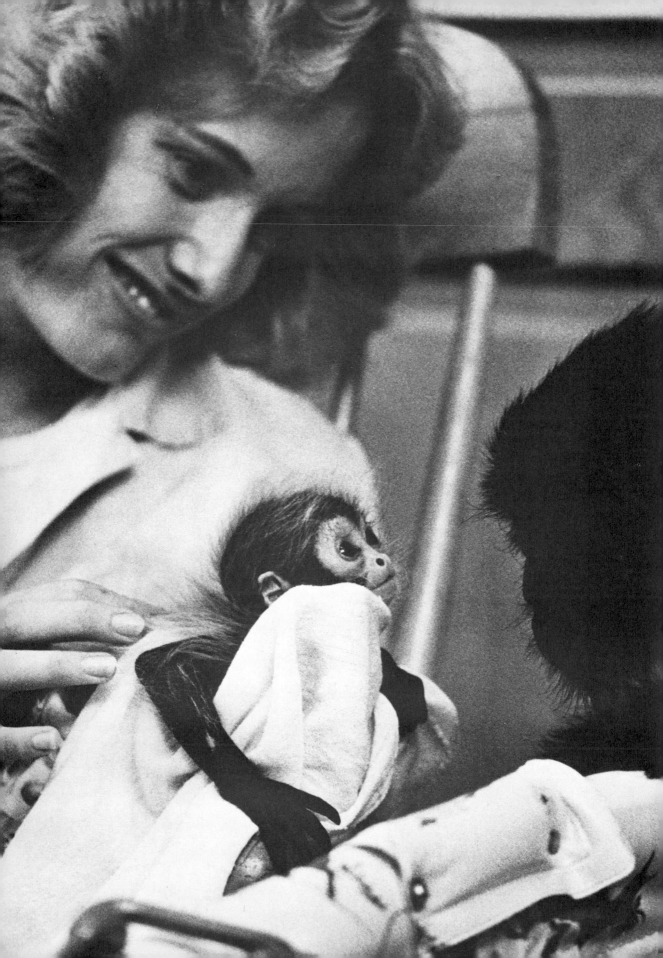

December 23

It's 7:00 P.M. The nursery attendant who takes over from Loretta at midnight has just called in sick. IF YOU'RE SICK, STAY HOME. That's the first thing every woman who's trained as a nursery attendant learns. If an attendant has even a sniffle or a mild sore throat, she cannot report to work. Kalind sleeps the whole night through after his 11:00 P.M. feeding, but Drainola, Lock, and Lisa all get an additional bottle around 3:00 in the morning, and will need someone to look after them. Tacked on the wall right above the telephone is a listing of veterinarians who are on standby. Right next to this is a list of nursery attendants that includes three children's zoo attendants who have been trained to work in the nursery. Loretta tries to reach one of the standby attendants, without any success. One woman has a bad cold, and no one answers at the other numbers.

It's 7:15 P.M., time for Drainola's feeding. Kalind is practically climbing out of his playpen to get a better look at the tiny monkey, so Loretta moves her rocking chair right next to his playpen before settling down with Drainola.

Drainola has gained three ounces, one ounce for each week she's been in the nursery. Long wisps of golden hair are beginning to show up against her black fur. When she's grown, she'll have lost most of her black fur. Her new coat will be a colorful golden-brown.

"Is Kalind your brother, your big brother? Don't be frightened; it's all right," Loretta tells Drainola soothingly. "You know what? I think he thinks you're sweet."

30

Holding the small monkey in one arm, Loretta reaches into Drainola's crib and gathers two baby blankets together, making a little nest for Drainola to lie in so she will feel more secure in the vastness of her crib.

By 10:00 P.M. Loretta has not been able to find anyone to take the midnight to 3:30 A.M. shift. Although she's tired, she's sure she'll be able to stay awake. She decides not to bother any of the regular attendants. She'll stay over.

January 10

Three-and-a-half-month-old Lisa is happy and contented lying in her crib, playing with her hands and feet. But Lock realizes that there's a whole new world beyond his crib. He's attuned to all the nursery noises. There's something going on that sounds interesting. He rolls and pulls himself over to the side of his crib and, sucking his fingers, he stays with his face pressed against the bars, peering out.

There's an aardvark in the nursery! Just two days old, she opened her eyes this evening, shortly before dusk. "There's something beautiful about a squirming, breathing, helpless, newborn infant. And sometimes, while you're watching, the infant starts to open its eyes. It makes me catch my breath. It's always," Loretta says, "a special moment."

Pushed out of the burrow by her mother, the pinkish gray, four-pound, four-ounce infant will be raised in the nursery. Adult aardvarks at the San Diego Zoo live in dark underground enclosures that are much like the burrows aardvarks

dig for themselves in the wild. But even though they live in similar environments, mother aardvarks in captivity almost always desert their newborn babies. No one knows what causes this strange behavior.

A-Seven, as the nursery attendants have named her, is the nursery's seventh aardvark. Some of the other babies have been named Wrinkles, Crinkles, Minerva, and Whatchama-callit. "You get pretty giddy thinking up aardvark names," Loretta says. "But I never pay much attention. To me all aardvarks are Piggy."

Large-eared, rough, wrinkly-skinned A-Seven, with her blunt-tipped muzzle, does look something like a pig. Aard-varks come from southern Africa and, in Afrikaans, a South African dialect, *aardvark* means "earth pig." Fossil records in-dicate that these strange-looking mammals were living some fifteen million years ago.

It's past 9:00 P.M., but no one is asleep. Though A-Seven barely makes a sound, all the creatures know by now — there's a new animal in the nursery. Drainola lifts up her head trying to see. Lisa slowly, determinedly, pushes and pulls her-

32

self over to the side of her crib to look out. Kalind can't seem to make up his mind what he thinks of the aardvark. Curious, Lock pulls himself up to get a better look. After a while he falls down, lands in an awkward position on his back, and makes short, rasping, loud "I'm in dire trouble" screams, till he's picked up and comforted.

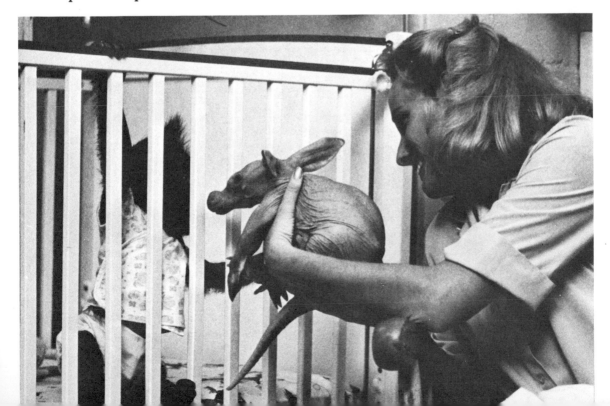

January 14

Though Lock and Lisa are close to four months old, they are still very helpless. But they now make their wants and needs clearly known. They are becoming very definite personalities. Lisa is cautious. She doesn't try to do anything till she's able to do it. She's less demanding than Lock. Lock never hesitates. He tries to do everything he wants, and he often succeeds. When Lock wants attention, he wants it the minute he wants it.

Loretta watches Lisa, who is sitting up holding on to the bright green plastic chain that's strung across the top of her crib. "Come on," Loretta says to her, "you've been able to do that for days. Aren't you ready to climb? You can do everything your brother can do; you're just as strong. You're a big pretend, that's all," she tells the little orang as she picks her up to feed her.

Lisa is not at all interested in her bottle. Loretta opens the

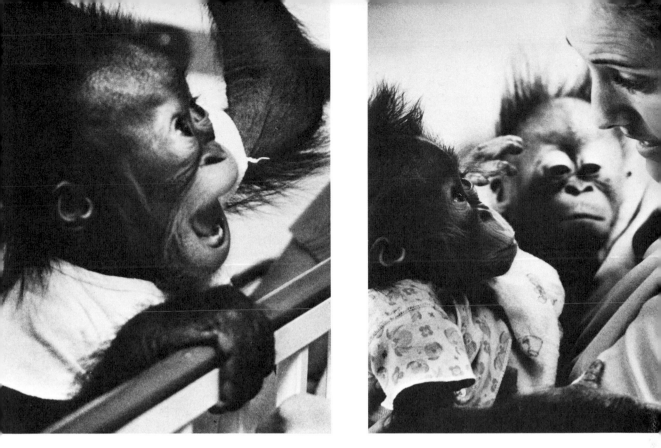

baby's mouth, looks at, then runs a finger along, her gums. "Teething," she pronounces. She massages Lisa's gums gently with her finger and says, "Well, I guess you're going to do something before Lock."

At 4:30 P.M. Loretta writes in the record book, "Lisa's two front, bottom teeth are starting to come in."

January 25

Everything in the nursery has an exact place. Terry bath towels, bright blue, orange, yellow, and lavender, are neatly folded to the same exact size and stacked on the top cabinet shelf. So are the cotton baby blankets. White cloth diapers and rubber pants, nail clippers and some medical supplies are on the shelf below. Bath supplies and a baby scale are stored in a bottom cabinet that's right next to a small refrigerator.

Each morning a complete day and night's supply of each animal's formula is made up. The ape and monkey formulas are similar to the formula a human baby would drink. A-Seven's formula is a combination of pig's milk, meat extract, minerals, and vitamins. The formulas are poured into neatly labeled bottles and stored in the refrigerator.

Loretta reaches into the refrigerator and gets out A-Seven's bottle. Though A-Seven will soon start growing fast, and looks very sturdy, young baby aardvarks are exceedingly delicate. They are susceptible to chills; a sudden drop in temperature can be very dangerous, and it is often difficult to get a baby to drink from its bottle. Minerva aardvark, who was in the nursery a few years ago, had to be force-fed steadily for a month before she began drinking voluntarily.

Two-week-old A-Seven is doing nicely. She's just graduated from her incubator to a crib. She's fed every two hours. Sometimes she has to be force-fed, but most of the time she takes her bottle willingly. "Are you going to be good? Are you a good Piggy? Come on ... open ... open. That's my girl," Loretta says, "that's my Piggy." The little aardvark fits snugly into the palm of Loretta's hand. A-Seven is easy to manage because her thick claws are rubbery soft. As she grows her claws will harden. When she's grown, her strong, heavy claws will be a powerful digging tool. In the wild these shy, nocturnal creatures can dig their way back into their burrow with astonishing speed.

The feeding's over, but Loretta holds A-Seven awhile

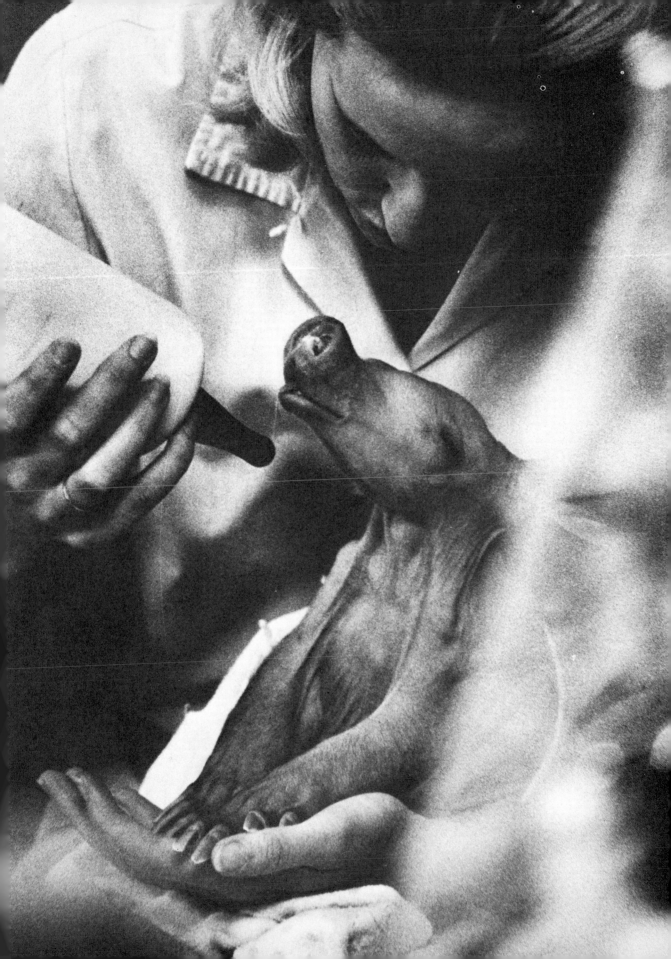

longer. "They only like to be held when they're very little," she explains, "and they're so cute." Then she adds, "I know not everyone thinks they're cute, but if a baby's mother doesn't, who will?"

January 30

It had been the kind of afternoon when all the animals needed attention at once. It was to be that kind of night too.

Loretta arrives shortly before 3:30 P.M. She checks with JoAnn to find out how the babies have been. She puts on a fresh nursery smock and is beginning to give Drainola her bottle, when she catches sight of Kalind. He is not watching Drainola, he is not playing with his toys, he is looking over at Lock's crib with an amazed expression on his face. Loretta turns and sees Lock, his feet balanced on top of his bumper pad, hanging precariously over the edge of his crib. Still holding Drainola, she dashes over, reaching Lock just before he nose-dives. "No," she says, hugging the little orang. Lock, not at all concerned about his close call or Loretta's disapproving tone of voice, thinks it's very nice to be picked up. He fastens his hold tightly around her neck. Normally quiet Drainola lets out a piercing scream. Loretta lowers her gently down into her

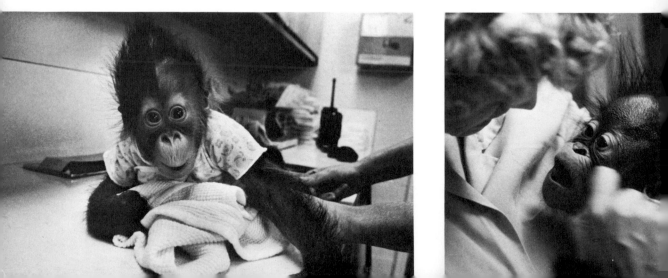

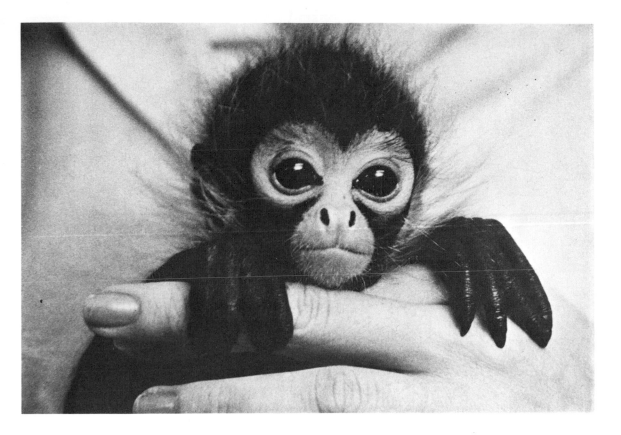

crib. She whisks Lock's bumper pads out of his crib, pries his fingers apart, and sets him down in his crib. She raises the sides up one notch, to their highest level. Then she takes the pads into the storeroom and puts them on a top shelf.

"Did Lock frighten you? He didn't mean to. You're okay," Loretta tells Drainola. The baby monkey sucks away on her bottle. "Hee hee hee," Kalind chatters excitedly. He's thoroughly enjoyed Lock's escapade.

By 4:15 P.M. Lock has discovered that he can't chin himself or climb over the top of his crib anymore. He shrieks loudly. Loretta ignores him and feeds and changes Lisa. She waits till Lock has stopped making a racket before giving him his bottle. Kalind is rocking back and forth in his playpen. "What's

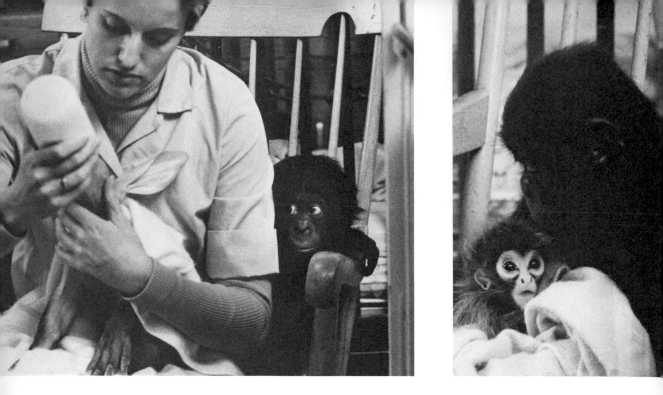

the matter, BooBoo?" Loretta asks. Like a mother pygmy chimp, she tickles him rather roughly, around his stomach and on his sides. "Is everyone getting attention except you? Do you feel like being part of everything that's going on?" Loretta asks. She picks Kalind up and sets him in the rocking chair, then brings A-Seven over for her 5:15 P.M. feeding. A-Seven refuses to take her bottle. Loretta knows that she's been doing that all day. "Silly Piggy," she says. Loretta shoves the bottle in the infant's mouth, clamps her hand around the aardvark's muzzle to keep it shut, then tips the aardvark's head way back, forcing her to drink. Then at 5:50 P.M. Kalind doesn't want his cereal. He practices all his delaying tactics — clamps his mouth shut, reaches for the spoon — but finally he finishes and is rewarded with his bottle. Loretta puts him back down in his playpen, then reaches in for a plastic bucket. She sweeps together most of the toys that are scattered over the playpen floor and puts them into the bucket, so Kalind can have the fun of pulling them out all over again.

Evening turns into night. Loretta comforts, scolds, plays, hugs, and cares for her nursery charges. Drainola and A-Seven get a bottle every two hours, Lisa and Lock every three, and Kalind every four. Lock wakes up. He tries climbing out of his crib again and finds out that he can't. He screams and wakes up Lisa, who also starts to scream. A-Seven gets another feeding. Her skin is getting very dry; it's cracking. Loretta rubs the aardvark down with a vitamin A and D oil. It's a cold night; you can feel it, even in the 78-degree nursery. Loretta puts an extra blanket over Drainola.

It's almost 11:30. Loretta is giving A-Seven her fourth bottle. The late night attendant will soon take over.

February 7

Somehow, Drainola caught a cold six days ago. Pretty soon all the nursery babies, except A-Seven, had one too.

The sniffly youngsters were examined by a veterinarian daily. The vets took their temperatures, looked down their throats and in their ears. They listened to their heartbeats, thumped them on their backs, and took blood samples. Kalind, Lock, and Lisa ran a slight fever of just under one hundred degrees with their colds. Normal body temperature for apes is about 98.6, just like people. Everyone was worried about the possibility of the babies' developing pneumonia, but yesterday, just five days later, the whole crew was well again.

It's 4:40 P.M. Veterinarian Jane Meyers comes into the nursery to give the animals a follow-up check. Kalind, who was very good about being examined when he was sick, now discovers something — he doesn't like vets after all. He winces when the cold stethoscope touches his chest, and clings to Loretta. Back in his playpen he seems to have forgotten all about the vet, but when she starts examining Drainola, his soft fur bristles, he leans forward, bares his teeth menacingly, and makes threatening *hee hee* sounds. Even though he's still a baby, he's not going to let anyone hurt his Drainola.

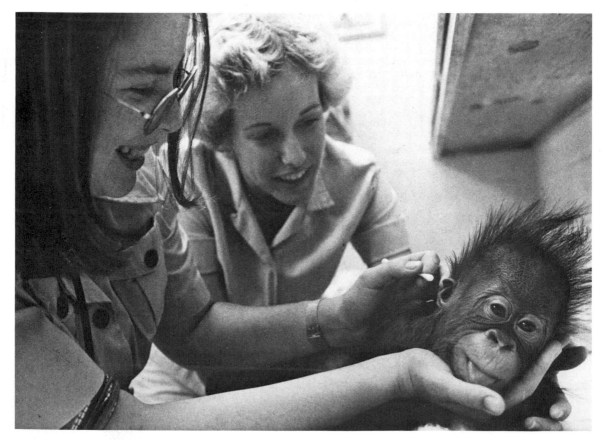

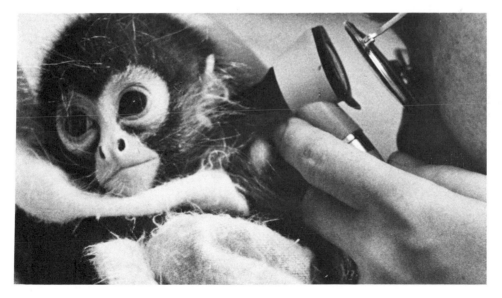

February 9

Drainola is now almost two and a half months old, and she's eating solids. Loretta feeds her strained applesauce-cereal. The spider monkey takes each spoonful very slowly. She seems to like it.

The twins are four and a half months old. Lisa has four teeth. Lock's first teeth are just starting to come in and he isn't acting at all like his normal bouncy self. He whines, wants to be picked up, then wants to be put down, and he doesn't feel much like taking his bottle.

Ten-month-old Kalind climbed out of his playpen earlier

today. But now, at 7:00 P.M., he's back in his pen, leaning out, watching one-month-old A-Seven get her bottle. He doesn't seem at all interested in climbing out again.

Kalind knows right from wrong, and he understands the nursery rules. Unlike orangutans, who are solitary animals in the wild, pygmy chimpanzees are by nature social animals. They like to get along. Of course, there are times a pygmy chimp will do what he knows is wrong, especially if he thinks he can get away with it. When Kalind was just five months old and was teething, he realized that if he waited till no one was in the nursery, he could move the bumper pads in his crib and chew all along the bottom of his crib bars. Then, by putting the pads back into place, he could cover the damage. His secret was discovered when one of the attendants was straightening his crib.

Now, however, Kalind seems less interested in mischief, or getting his own way, than he was a few months ago. He's more interested in the nursery and everything that goes on in it. He knows the daily routine and seems to enjoy it. He notices if anything is different, or out of place, and he doesn't like it. He knows how things should be. He likes being praised, and he's easy to discipline. "He is," Loretta says, "nice to be around."

Though Kalind still plays contentedly in his playpen, the attendants and vets agree it's time for the growing pygmy to be in a larger enclosure, so he can practice his climbing and swinging. The maintenance department at the zoo is constructing a big cage for Kalind and fitting it with special equipment. It will be moved into the nursery soon.

47

February 13

It took Kalind a whole day to get used to the new cage. The pad on its floor was his old playpen pad. The toys were all his and he did seem to understand that it was his new place. He held on to his blanket for the first couple of hours, played with his toys a little, and did a lot of looking out. After he was fed, he didn't mind going back in. But Loretta was afraid that if he woke up in the cage in the middle of the night he might find his new surroundings frightening. So after his last feeding, instead of returning him to his cage, she put him down for the night in his old baby crib.

Kalind went back into his cage first thing in the morning. It didn't take him long to discover he loved it! He swung from the chain, first with both hands, then with one. He practiced striking strong, adult male, "I'm a big chimp now" poses. He seemed a thoroughly happy pygmy chimp.

February 22

"The animals get bathed every three or four days," Loretta explains, "but it's usually done late morning. Just sometimes, someone doesn't get to it and then I bathe them. It's fun."

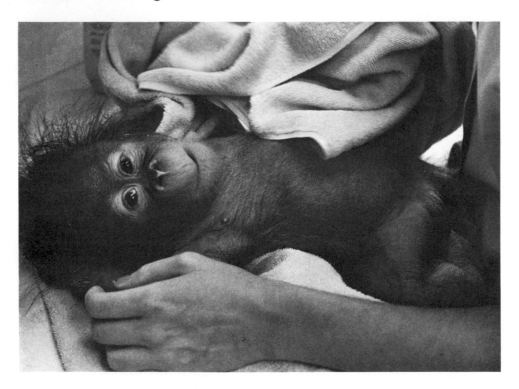

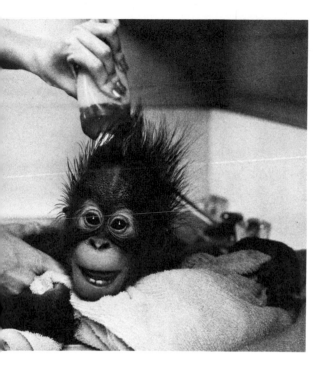
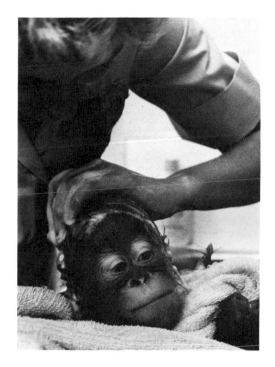
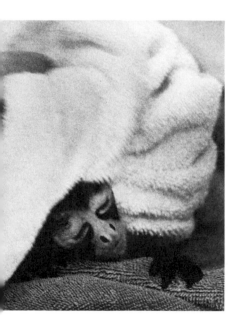
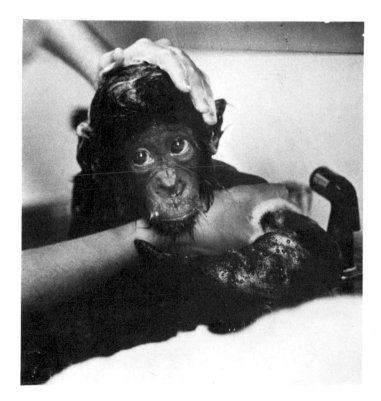

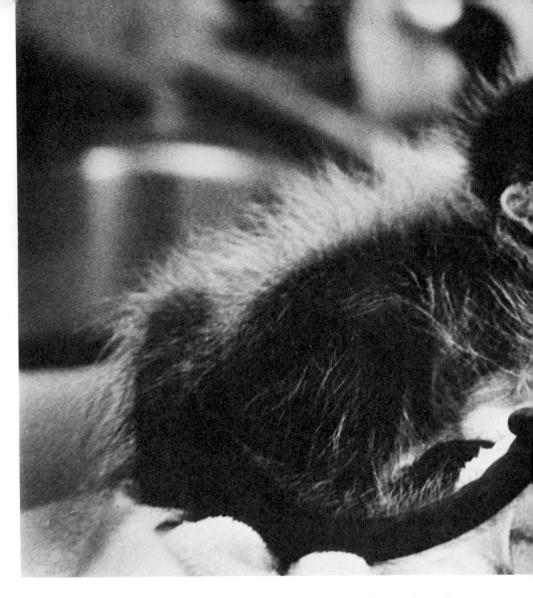

Loretta sudses and scrubs the protesting, sometimes howling babies. Lock, after the first yelp, doesn't seem to mind it, and he loves the rubdown afterward. He's ticklish. Lisa grabs for a towel to hold on to when she feels herself in deep water. Drainola seems to have an "I'll just have to put up with it" attitude. "You're good," Loretta tells her, "you're the best one." Kalind tries drinking the soapy water. A-Seven is the only one who doesn't get bathed, but she does get wiped down with a warm, damp washcloth, then dried and oiled.

At 6:00 P.M., Kalind finds his ears are perfect pockets for unwanted mashed bananas. "No, no," Loretta tells him sternly, trying to keep a straight face, "bananas do not go in your ears!"

From 10:00 P.M. till midnight Loretta works nonstop, feeding, changing, and getting everybody ready for bed. All the growing babies now sleep through the night, so from the time Loretta leaves, no one comes into the nursery till 6:30 in the morning.

March 1

Lock and Lisa are over five months old. They made the big move to a cage a few days ago. It's a large double cage. They weren't at all frightened. They were used to the idea. They'd been watching Kalind in his cage, playing and showing off. The first few days, Lock pushed and trampled on Lisa. He didn't seem aware that she had any feelings. Then Lisa learned to push back.

Now suddenly, Lock is fascinated by mouths. One of his favorite games is prying open Lisa's mouth and examining the inside. Lock also likes to climb up the sides of his cage, but he hasn't figured out how to get down. Before Loretta leaves each night she mittens Lock's hands to keep him from climbing. She puts Lisa back into her crib, to protect her from Lock's rough play. If the twins were living in the wild, they'd be with their mother all day, holding on to her back, or clinging to her stomach, as she moved about. At night, high up in a treetop nest, they'd sleep with their mother, snuggled against her for warmth and protection.

"Orangutans," Loretta says, "are expert mischief-makers. They can and do take everything apart." It's difficult for scientists who study apes to tell how smart orangutans are in comparison to gorillas or chimpanzees. They'll devise a test to see if an ape can learn how to use a tool to solve a particular problem, but the orang will become more interested in taking the tool apart than in solving the problem. There's a very old orang in the San Diego Zoo named Bob. Bob, who came to the zoo when he was about three years old, was a super escape artist. He managed in less than a month to escape from three escape-proof cages. Each time Bob was found, he would happily hold out his arms to his keeper, asking to be picked up. He only liked working his way out of cages; he never tried to get away.

Another orangutan who was in the nursery a couple of years ago took the roof off his cage. Loretta says he must have loosened or taken out forty or fifty screws that were holding the top in place. He didn't climb out of his cage. He was just having fun moving the top up and down when the attendant discovered him.

March 5

Eleven-month-old Kalind is not afraid to be out in the nursery by himself. "The whole nursery is getting to be his cage," Loretta says.

Scientists have observed chimpanzees in the wild for many years. They are just beginning to study pygmy chimpanzees in the wild. They've discovered that, like chimpanzees, they live in colonies and the family unit seems to consist of a mother and her children. The father probably associates with other grown males, and is not part of the family. Unlike lower, less intelligent animals, pygmy chimps can change their behavior. They can adapt to new surroundings. In the zoo, Kalind's father, mother, uncle, brother, and four sisters all live together in a large enclosure. Mother and father show equal concern and affection for their children and the whole family seems very close. They don't get angry often, but when they do, they get over it very quickly.

Kalind's brother and sisters also spent some time in the nursery and all of them were greeted joyously upon their return to the family. Loretta says, "When Kalind goes back, his mother will probably be the first one to recognize him. She'll pick him up and hold and cradle him as if he were a newborn. The children, chattering excitedly, will gather round. After a while the father will take charge. Once," Loretta continues, "when one of Kalind's sisters was returned, the father held the baby in his arms for so long, the keeper wondered if the baby would ever have a chance to eat."

Kalind is playing on the nursery floor. Loretta sets A-Seven down on the floor, and she starts running about. Kalind seems fascinated. "Go on, Bugaloo," Loretta says, coining a new nickname for Kalind, "get the Piggy." Kalind starts to chase the aardvark several times, but each time he stops. He's very curious, but not quite sure he wants to chase an aardvark.

A-Seven, who is almost two months old, weighs twelve pounds. She's gained five ounces since yesterday. "You're gaining weight very fast, even for a Piggy," Loretta tells the aardvark as she scoops her up. Loretta clips the aardvark's nails, then returns her to her crib.

A crib is not a good place for a growing aardvark who likes to dig, so a special nursery pen is being constructed for her. It will have a dirt floor for her to dig in, and will be far enough off the floor to keep A-Seven out of drafts. It will have Plexiglas sides so she can see out. It will also have to be very strong, for A-Seven will be five or six months old by the time she leaves the nursery and she'll weigh fifty pounds or more. If A-Seven grows up to be an average-size aardvark, she'll weigh one hundred and fifty pounds.

March 21

"I think," Loretta says, "if one of the orangs were taken out of the nursery for a few days, the remaining orang would be very lonely. It's hard to tell for sure. It's just a feeling. They don't play with each other yet. But they do grab hold of each other and," Loretta smiles, "sit on each other, and I think they get comfort from each other's physical presence."

Most of the San Diego Zoo's orangutans live in a large, sloping, grass-covered area. A huge multilevel climbing structure made out of eucalyptus logs is the focal point of activity for adult orangs and the older children. The younger orangs spend much of their time on the ground playing all kinds of pushing, pulling, and tumbling games. Their behavior often amazes people who know a lot about orangutans, because orangs in the wild show no interest in group living. Orangutans are plant eaters. They are very large eaters and spend

most of their day eating and searching for food. Some scientists think that orangs in the wild must stake out their own territory for foraging and not let anyone else in, in order to survive.

The zoo's orangutans are active and seem very happy. Occasionally a two-hundred-and-fifty-pound adult male orang will slap down, but not harm, a rambunctious older child. Most of the time, oldsters, youngsters, adults and children, males and females all get along together.

The twins are now six months old. In another nine to twelve months, they will be too big and active for the nursery. In addition, orangutan babies do not have a built-in desire to please. An orang mother in the wild, probably because she leads such a lonely life, does not discipline her children or teach them proper orang behavior. "No matter how loving orang babies are," Loretta says, "they do not abide by rules; they do what they like."

When Lock and Lisa are old enough to care for themselves (fourteen or sixteen months of age), they will join the zoo's other orangutans. Their mother will not accept them as her babies, but like most of the adult female orangs, she will act in a kindly and protective manner. The little orangs will be delighted to have two new playmates. Lisa and Lock will be shy, a little confused, but it won't take them long to feel that they belong.

Loretta and Drainola visit the orangs. Lock isn't at all interested in Drainola, but Lisa finds her fascinating.

It's 4:30 P.M., and time for Drainola's feeding. The baby spider monkey eats twelve spoonfuls of her fruit-cereal combination. Loretta praises her for eating so well, then gives her her bottle. Drainola, three months and three weeks old, now weighs one pound, nine and a half ounces. She weighs eight and a half ounces more than she did when she was born. She gets solids and two ounces of formula every three hours, starting at 7:30 in the morning and ending at 10:30 at night.

Like Lock, Lisa, and Kalind, Drainola now has a cage. It seems strange to see such a tiny baby in a huge cage. But Drainola is starting to climb. Spider monkeys have very long limbs and tails compared to the size of their bodies. It takes them quite a while till they can coordinate their legs and arms. Until they can, they're very awkward. But little by little they become more adept. A full-grown spider monkey can move with astonishing speed and can jump a chasm thirty feet wide. In a couple of months Drainola will be swinging and climbing the height and width of her cage. Then, in a few more months, when she is swinging and climbing with ease and is big enough and strong enough to take care of herself, Drainola will leave the nursery and join the zoo's other spider monkeys.

Loretta doesn't like to think about the babies growing up and leaving the nursery. But she knows she is raising the babies so that one day they'll all be big enough and strong enough to be on their own. "It's particularly hard," Loretta says, "to see the apes go. They're in the nursery so long, and I become so fond of them."

It takes Loretta a little longer than usual to clean off the nursery counter, because Kalind's hanging on to her back. "Get down, Bugaloo," she says sternly to the pygmy chimp. She gives him a little hug before putting him back in his cage. In the past, several of the pygmy chimps that were taken from the nursery when they were a year old landed back in the nursery a few months later with severe colds. So Kalind will stay in the nursery till he's almost two years old, or for another eleven or twelve months.

It's 6:30 P.M. The zoo is closed. The thousands of people who visit the zoo each day have left. The keepers have finished feeding the animals; they too are gone. It's almost dark. A chill wind blows across the zoo. There's a light on in the nursery.

Inside the cheerful yellow nursery it's cozy and warm. Kalind has just finished his cereal. Loretta starts to give him his bottle. He reaches up, touches her face gently, and begins to drink. Loretta talks to the almost-one-year-old pygmy chimp. She tells him how cute he is, how smart, and that she knows he's a little mischief-maker, but she loves him anyway.